Liberty Dendron

Above The Earth

by

Liberty Dendron

Above The Earth

© 2013 by L. A. Johnson Jr.

All rights reserved. No part of this book may be reproduced, stored in a retrieval system or transmitted in any form or by any means without prior written permission of the author, except by a reviewer who may quote brief passages in a review to be printed in a newspaper, magazine or journal.

First Edit by Janicea Johnson

First printing

All characters in this work are fictitious. Any resemblance to real persons, living or dead, is purely coincidental.

Mamba Books
"Where Knowledge is Wealth"

ISBN:

Published by: Mamba Books & Publishing

Dendron Virginia

Printed in the United States of America

Liberty Dendron

The sun often feels too hot, or shines too bright, without it the earth would be cold and dark.

Above The Earth

What Is The Solar System?

The Solar System is made up of all the planets that orbit our Sun. In addition to planets, the Solar System also consists of moons, comets, asteroids, minor planets, and dust and gas.

Everything in the Solar System orbits or revolves around the Sun. The Sun contains around 98% of all the material in the Solar System. The larger an object is, the more gravity it has. Because the Sun is so large, its powerful gravity attracts all the other objects in the Solar System towards it. At the same time, these objects, which are moving very rapidly, try to fly away from the Sun, outward into the emptiness of outer space. The result of the planets trying to fly away, at the same time that the Sun is trying to pull them inward is that they become trapped half-way in between. Balanced between flying towards the Sun, and escaping into space, they spend eternity orbiting around their parent star.

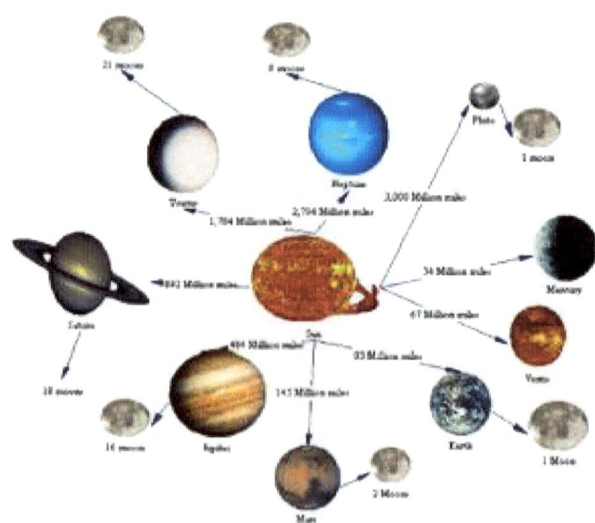

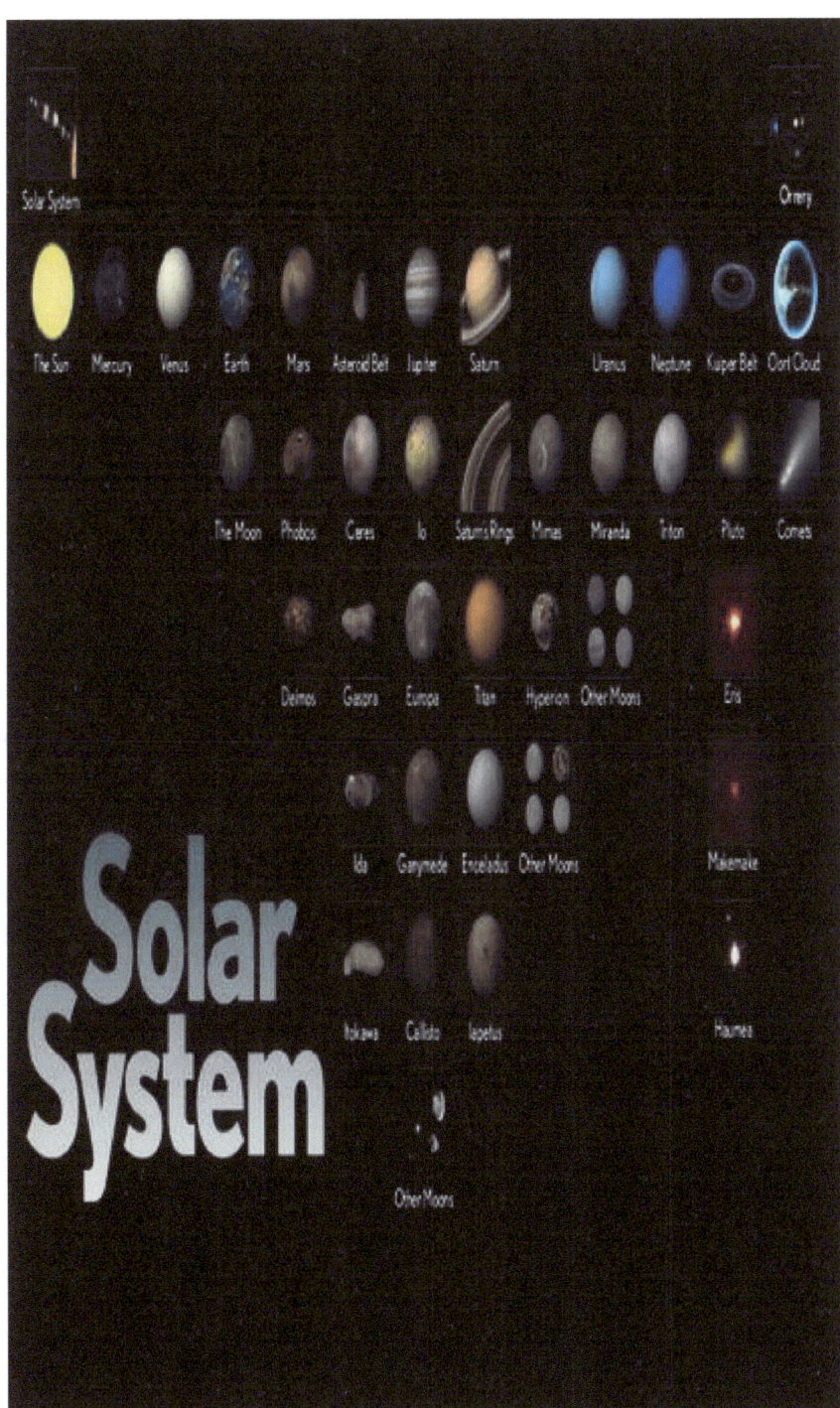

Above The Earth

How Did The Solar System form?

The creation of our Solar System took place billions of years before there were any people around to witness it. Our own evolution is tied closely to the evolution of the Solar System. Thus, without understanding from where the Solar System came from, it is difficult to comprehend how mankind came to be.

Scientists believe that the Solar System evolved from a giant cloud of dust and gas. They believe that this dust and gas began to collapse under the weight of its own gravity. As it did so, the matter contained within this could begin moving in a giant circle, much like the water in a drain moves around the center of the drain in a circle.

At the center of this spinning cloud, a small star began to form. This star grew larger and larger as it collected more and more of the dust and gas that collapsed into it.

Further away from the center of this mass where the star was forming, there were smaller clumps of dust and gas that were also collapsing. The star in the center eventually ignited forming our Sun, while the smaller clumps became the planets, minor planets, moons, comets, and asteroids.

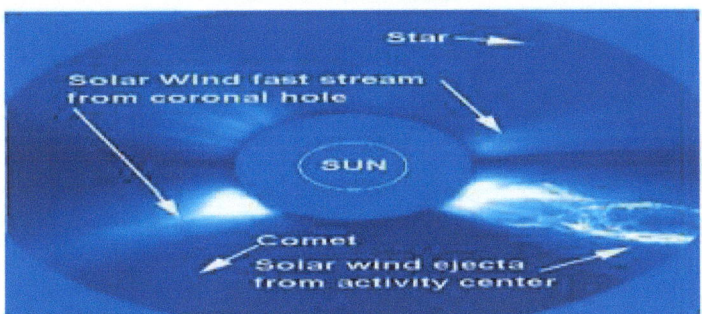

A Great Storm

Once ignited, the Sun's powerful solar winds began to blow. These winds, which are made up of atomic particles being blown outward from the Sun, slowly pushed the remaining gas and dust out of the Solar System.

With no more gas or dust, the planets, minor planets, moons, comets, and asteroids stopped growing. You may have noticed that the four inner planets are much smaller than the four outer planets. Why is that?

Because the inner planets are much closer to the Sun, they are located where the solar winds are stronger. As a result, the dust and gas from the inner Solar System was blown away much more quickly than it was from the outer Solar System. This gave the planets of the inner Solar System less time to grow.

Another important difference is that the outer planets are largely made of gas and water, while the inner planets are made up almost entirely of rock and dust. This is also a result of solar winds. As the outer planets grew larger, their gravity had time to accumulate massive amounts of gas, water, as well as dust.

Liberty Dendron

Bizarre weather is not restricted to Earth. Hurricanes on Earth are merly specks of dust compared to some of the cataclysms currently taking place around the solar system. Jupiter, for example, is going through a tumultuous time right now. The gas giant has suffered more meteor impacts in the past years than has ever been observed, and large cloud formations are spontaneously changing color or disappearing as quickly as they form.
But Jupiter is not the only planet in our solar system that experiences bizarre weather. Icy methane rainstorms, planet-wide sand storms, and lead-melting temperatures afflict other planets and their moons.

Beyond The Oort Cloud

The Sun's solar winds continue pushing outward until they finally begin to mix into the interstellar medium, becoming lost with the winds from other stars. This creates a sort of bubble called the Heliosphere. Scientists define the boundaries of the Solar System as being the border of the Heliosphere, or at the place where the solar winds from the Sun mix with the winds from other stars.

The Heliosphere extends out from the Sun to a distance of about 15 billion miles, which is more than 160 times further from the Sun than is the Earth.

Above The Earth

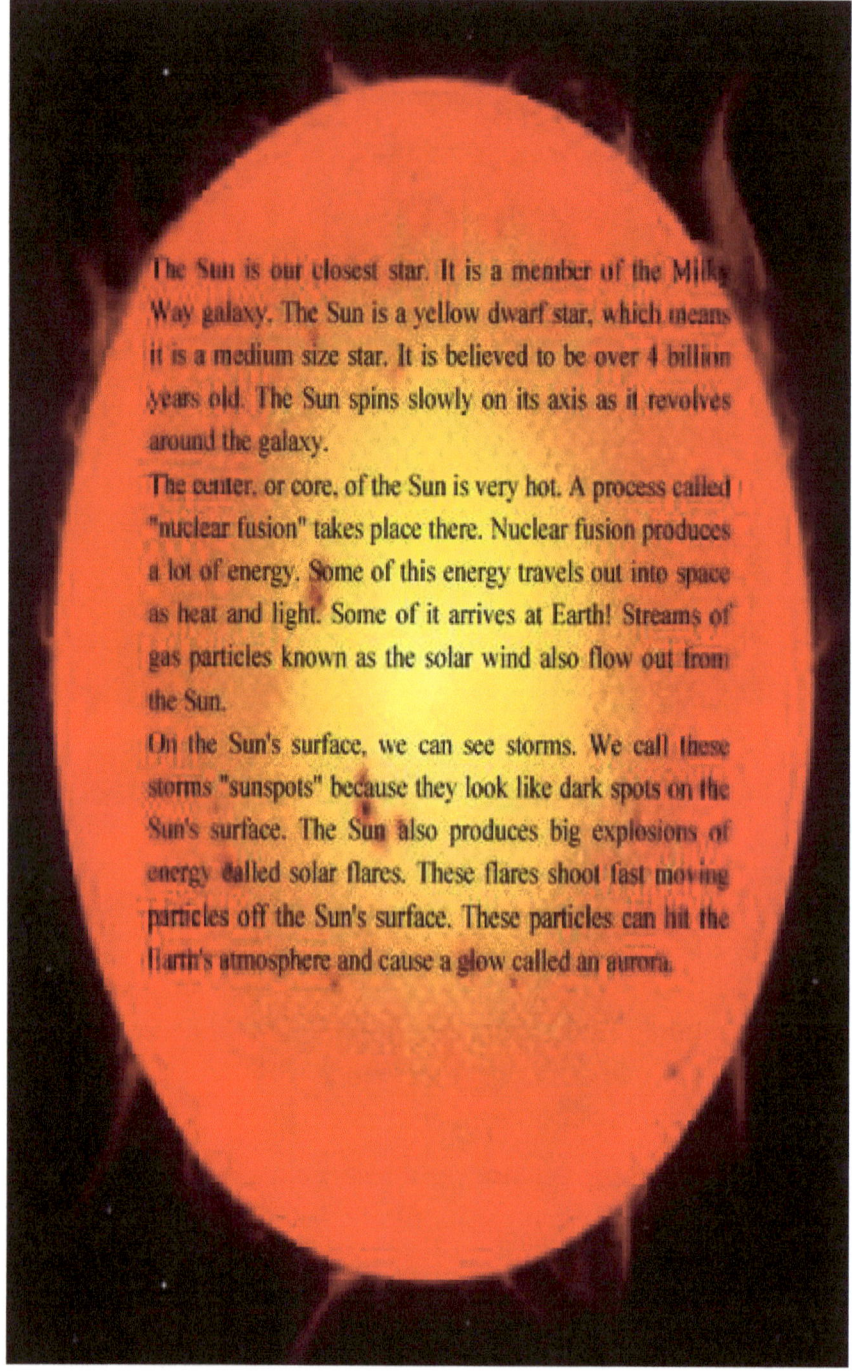

The Sun is our closest star. It is a member of the Milky Way galaxy. The Sun is a yellow dwarf star, which means it is a medium size star. It is believed to be over 4 billion years old. The Sun spins slowly on its axis as it revolves around the galaxy.

The center, or core, of the Sun is very hot. A process called "nuclear fusion" takes place there. Nuclear fusion produces a lot of energy. Some of this energy travels out into space as heat and light. Some of it arrives at Earth! Streams of gas particles known as the solar wind also flow out from the Sun.

On the Sun's surface, we can see storms. We call these storms "sunspots" because they look like dark spots on the Sun's surface. The Sun also produces big explosions of energy called solar flares. These flares shoot fast moving particles off the Sun's surface. These particles can hit the Earth's atmosphere and cause a glow called an aurora.

Liberty Dendron

Earth orbits around our Sun. Its orbit is nearly circular, so that the difference between Earth's farthest point from the Sun and its closest point is very small. It takes roughly 365 days for the Earth to go around the Sun once. The time it takes for the Earth to go around the Sun one full time is what we call a year. The combined effect of the Earth's orbital motion and the tilt of its rotation axis result in the seasons.

It takes one day, 24 hours to complete one rotation As Earth rotates, half of the Earth is always illuminated by the sun and half of the Earth is always dark. The sun gives us light, heat, and life. Its energy is absorbed buy plants, and plants make oxygen for us to breathe. It also gives us electricty, cooled by solar panels, and powers our cities and cars.

When the sun was a tiny star, it was sourounded by dest, gasses, ice, and everything the Earth, the moon, and everything that planets are made of. Slowly planrts formed.

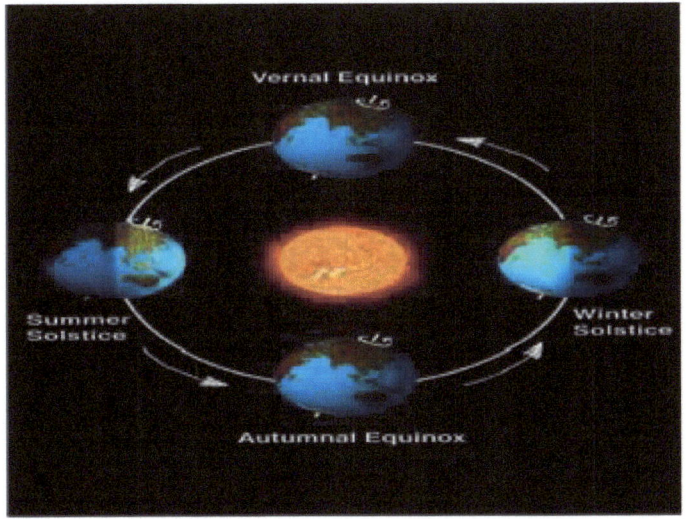

Above The Earth

And our solar system was born.

"With so much sunlight, is Earth the only place where there is life?" Perhaps not, but there are lots of planets, and alot of moons, and alot of the things that make a planet. There is alot of things like gas, dust, and ice. Some of them are expanding.

Above The Earth

Mercury is the planet closest to the Sun. It has very little atmosphere, its dusty surface of craters resembles the Moon. It has only three days every two years. It has very long and cold nights, and it is to-fifths the size of Earth in diameter; it is the second smallest planet in the solar system.

Liberty Dendron

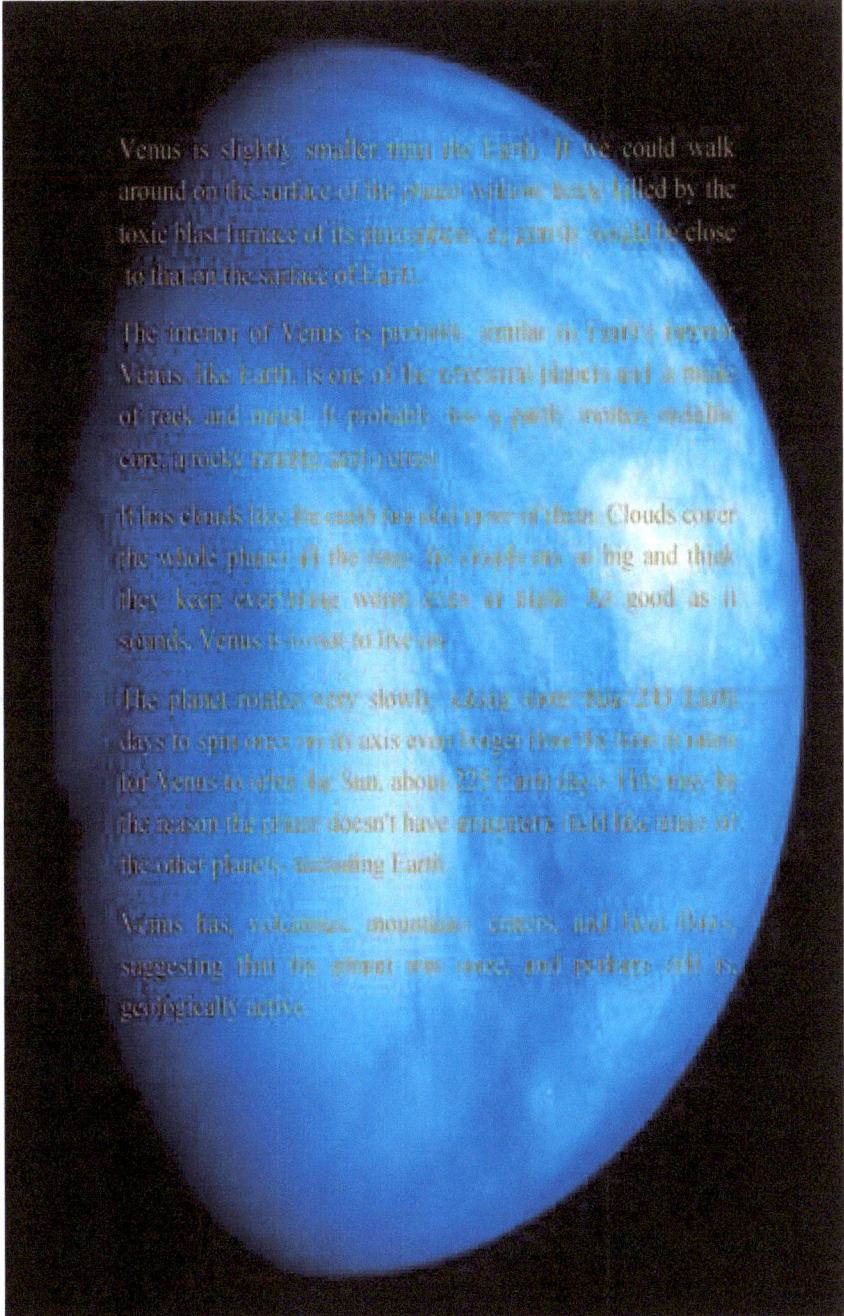

Venus is slightly smaller than the Earth. If we could walk around on the surface of the planet without being killed by the toxic blast furnace of its atmosphere, its gravity would be close to that on the surface of Earth.

The interior of Venus is probably similar in Earth's interior. Venus, like Earth, is one of the terrestrial planets and is made of rock and metal. It probably has a partly molten metallic core, a rocky mantle and a crust.

It has clouds like the earth has also types of them. Clouds cover the whole planet all the time. Its clouds are so big and thick they keep everything warm even at night. As good as it sounds, Venus is to hot to live on.

The planet rotates very slowly, taking more than 243 Earth days to spin once on its axis even longer than the time it takes for Venus to orbit the Sun, about 225 Earth days. This may be the reason the planet doesn't have anti-matters to kill like mars or the other planets, including Earth.

Venus has volcanoes, mountains, craters, and lava flows, suggesting that the planet was once, and perhaps still is, geologically active.

Above The Earth

The Red Planet as Mars is often called, is the fourth planet from the sun. In a lot of ways, Mars looks a lot like earth, though instead of blue oceans and green land, Mars is home to an ever present red tint. This is because of a mineral called iron oxide. iron oxide is very common on the planet's surface.

However, when you look past the surface differences, our planets are similar in a lot of ways. It is easy to forget that Earth is not the only planet in the solar system. Seven other planets whiz around the sun just like ours. Of those planets, none of them are closer or more engaging to the imagination than Mars.

Mars has both North and South polar ice caps, much like Earth. Also like Earth, both ice caps are made mostly of frozen water. With so much water frozen in the ice caps of Mars, some scientists think that life could have once existed there. Mars has seasons like Earth too. These seasons are much longer than Earth seasons because Mars is so much farther from the sun. Mars is cooler than the earth because its farther away. Water on Mars is frozen most of the time. Mars might have looked like earth a long time ago.

Jupiter is the fifth planet from the Sun, it is the largest planet in the Solar System. It is a gas giant. It is two and a half times the mass of all the other planets in the Solar System combined.

Jupiter is made out of thick hot gasses that spin and twist in giant sreeams. It even has its own solar system. It has dozens of moons, and, an enormous magnetic field that forms, a mini solar system.

Above The Earth

Some of Jupiters moons look alot like planits.

Jupiter has four large moons and dozens of smaller ones, there are about 67 known moons. Galileo first discovered the four largest moons of Jupiter, Io, Europa, Ganymede, and Callisto in 1610; these moons are known as the Galilean moons.

Io is a large, rocky, volcanically active moon of Jupiter. Its volcanoes spew out molten sulfur, making it a very colorful moon.

IO

EUROPA

Europa is a large, dense, icy moon of Jupiter. Europa is the smoothest object in our Solar System. Its surface is covered with long, crisscrossing trackways, but few craters, on water ice. Frozen sulfuric acid has been found on its surface. Europa's diameter is less than 2,000 miles (3,138 km), smaller than the Earth's moon.

Ganymede is the largest moon in the solar system; it is also larger than the planets Mercury and Pluto. It is the largest moon of Jupiter, a large, icy, outer moon that is scarred with impact craters and many parallel faults. It is an icy moon with a cold barren surface.

GANYMEDE

CALLISTO

Callisto is a large, icy, dark-colored, low-density outer moon of Jupiter, with impact craters. It has a diameter of about 3,000 miles. It is the second-largest moon of Jupiter; it is about the size of Mercury. Callisto has the largest-known impact crater in the Solar System, Valhalla, which has a bright patch across.

Saturn is the sixth planet from the sun, it is home to a vast array of intriguing and unique worlds. From the cloud-shrouded surface of Titan to crater-riddled Phoebe, each of Saturn's moons tells another piece of the story surrounding the Saturn system.

Sixty-two moons travel around Saturn. They come in a variety of sizes and compositions, from almost pure ice to rocky material, as well as a combination of both. Their journeys around the ringed planet average from half an Earth day to just over four Earth years. Saturn is not the only planet with rings, Jupiter, Neptune and Uranus have them too, but they are not as obvious as Saturn's.

Some of the moons travel inside the gaps of the rings, clearing paths through the debris. Others orbit farther out. Several of the moons interact with one another, affecting their orbits. Larger moons may trap smaller moons, keeping them nearby, but sixteen moons always face Saturn.

Saturn's moons formed early in the history of the solar system. Titan, makes up 96 percent of the mass orbiting the planet. Scientists think that the system may have originally housed two such moons, but the second broke up, and creating the debris that formed the rings, and smaller, inner moons. Another theory suggests that the system originally housed several large moons, similar to Jupiter's Moons. The violent collision could have scattered the debris that would have later drawn together into the smaller moons.

Titan's Surface

Titan is the only moon in the solar system known to have clouds and a mysterious, thick, planet-like atmosphere. Titan's atmospheric pressure is about 60 percent greater than the Earth.

It is a cloudy moon, just bigger than the planet mercury, its clouds don't keep it worm like Venus. Instead the keeps it cools. Titan looks allot like earth might have at the very beginning of life. It takes Titan almost 16 days to complete a full orbit of Saturn.

Researchers believe they have discovered clues that appeared to indicate that primitive aliens could be living on the moon. They have analyzed the complex chemistry on the surface of Titan, which experts say is the only moon around the planet to have a dense enough atmosphere. This suggest that life forms may or have been breathing in it's atmosphere or feeding on its surface; but astronomers say the moon is too cold to support even liquid water on its surface.

Black Holes

Black holes are one of the most mysterious and powerful forces in the universe. A black hole is where gravity has become so strong that nothing around it can escape, not even light. The mass of a black hole is so compact, so dense, that the force of gravity is too strong for even light to escape.

Black holes pull matter and even energy into themselves—but no more so than other stars or cosmic objects of similar mass. That means that a black hole with the mass of our own sun would not "suck" objects into it any more than our own sun does with its own gravitational pull.

Above The Earth

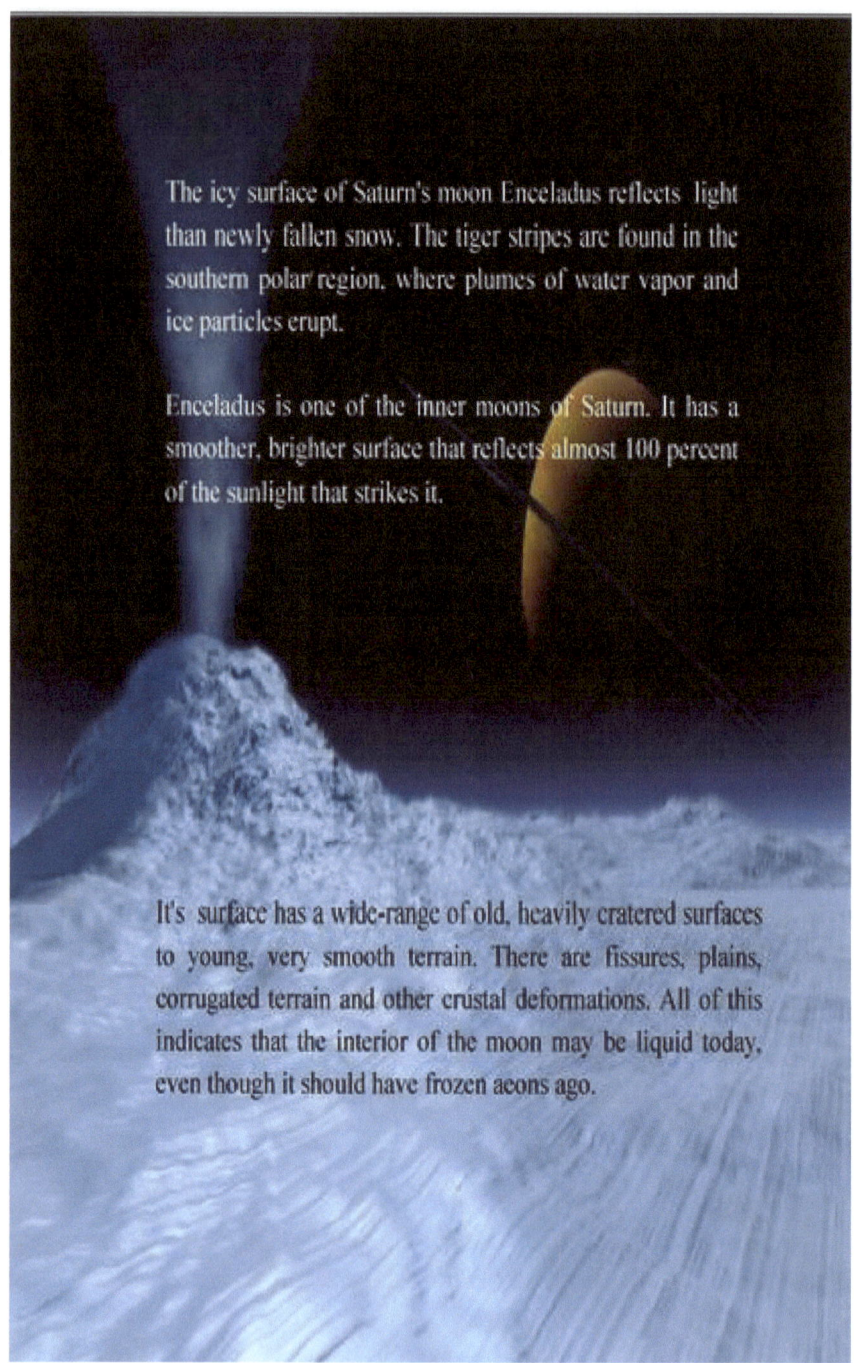

The icy surface of Saturn's moon Enceladus reflects light than newly fallen snow. The tiger stripes are found in the southern polar region, where plumes of water vapor and ice particles erupt.

Enceladus is one of the inner moons of Saturn. It has a smoother, brighter surface that reflects almost 100 percent of the sunlight that strikes it.

It's surface has a wide-range of old, heavily cratered surfaces to young, very smooth terrain. There are fissures, plains, corrugated terrain and other crustal deformations. All of this indicates that the interior of the moon may be liquid today, even though it should have frozen aeons ago.

Liberty Dendron

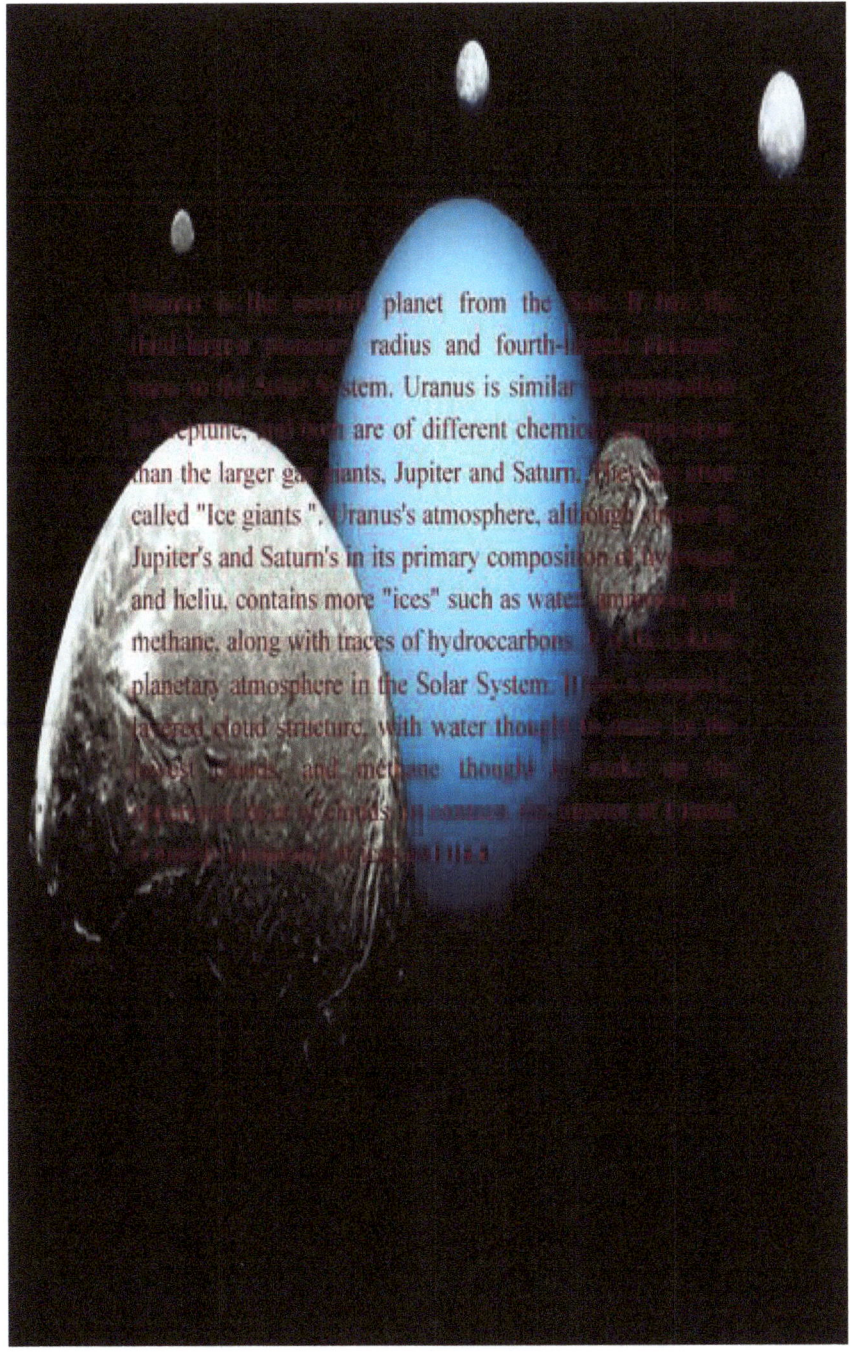

Uranus is the seventh planet from the Sun. It has the third-largest planetary radius and fourth-largest planetary mass in the Solar System. Uranus is similar in composition to Neptune, and both are of different chemical composition than the larger gas giants, Jupiter and Saturn. They are often called "Ice giants". Uranus's atmosphere, although similar to Jupiter's and Saturn's in its primary composition of hydrogen and helium, contains more "ices" such as water, ammonia, and methane, along with traces of hydrocarbons. It has the coldest planetary atmosphere in the Solar System. It has a complex, layered cloud structure, with water thought to make up the lowest clouds, and methane thought to make up the uppermost layer of clouds. In contrast, the interior of Uranus is mainly composed of ices and rock.

Above The Earth

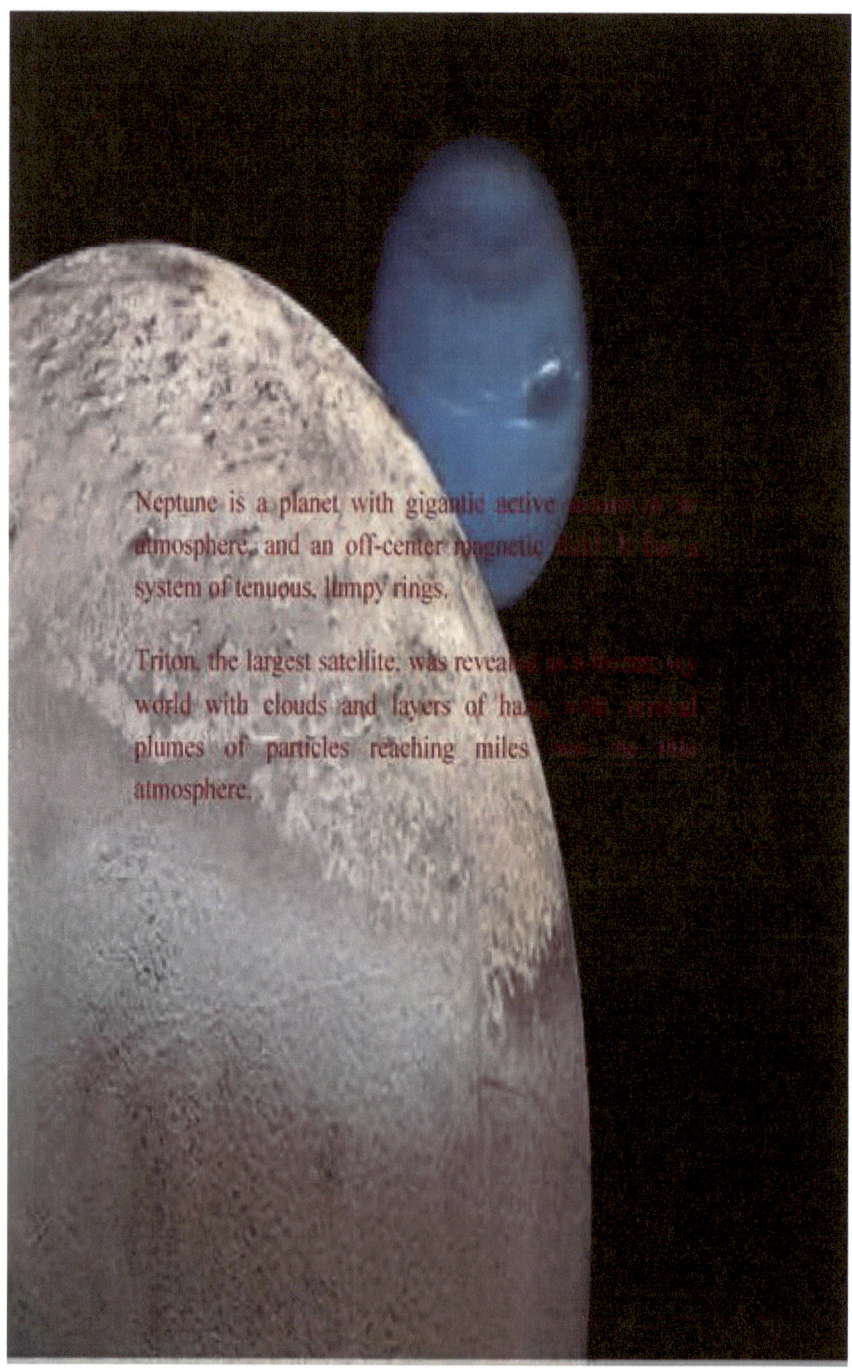

Neptune is a planet with gigantic active storms in its atmosphere, and an off-center magnetic field. It has a system of tenuous, lumpy rings.

Triton, the largest satellite, was revealed as a frozen icy world with clouds and layers of haze, with vertical plumes of particles reaching miles into the thin atmosphere.

Liberty Dendron

The Asteroid Belt is a region between the inner planets and outer planets where thousands of asteroids are found orbiting around the Sun. More than 7000 asteroids have been discovered. Several hundred more are discovered each year. There are undoubtedly hundreds of thousands more that are too small to be seen from the Earth.

The largest asteroid by far is Ceres. It is 914 km across and contains about 25% of the mass of all the asteroids combined. The next largest are Pallas, Vesta and Hygiea which are between 400 and 525 km across. All other known asteroids are less than 340 km across. Astroids are rarely visible with the naked eye, but many are visible with binoculars or small telescopes.

Origin

It is believed that, during the first million years of the solar system history, planets formed by accretion of planetesimals. Repeated collisions led to the familiar rocky planets and to the gas giants cores.

In this zone of the system the strong gravity of Jupiter inhibited the final stages, and the planetesimals could not form a single planet. The planetesimals instead continued to orbit the Sun as before. In this sense the asteroid belt can be considered a relic of the primitive Solar System, but many observations point to an active evolution of the physical conditions so the asteroids themselves are not particularly pristine. Instead, the objects in the outer Kuiper belt are believed to have had little change since the solar system's formation.

Asteroid belt environment

Despite popular imagery, the asteroid belt is mostly empty. The asteroids are spread over such a large volume that it would be highly improbable to reach an asteroid without aiming carefully.

Nonetheless, tens of thousands of asteroids are currently known, and estimates of the total number are in millions range. About 220 of them are larger than 100 km. The biggest asteroid belt member is Ceres, which is about 1000 km across. The total mass of the Asteroid belt is estimated to be 2.3×1021 kilograms (of which more than a third is accounted for by Ceres), which is less than that of Pluto.

The high population makes for a very active environment, where collisions between asteroids occur very often in astronomical terms. A collision can breakan asteroid into many small pieces leading to the formation of a new asteroid family, or may glue two asteroids together if it occurs at low relative speeds. After five billion years, the current Asteroid belt population bears little resemblance to the original one.

Asteroid belts are a staple of science fiction stories less concerned with realism than with drama, since they are always portrayed as being so dense that adventurous measures must be taken to avoid an impact. Proto-planets in the process of formation may look like that, but asteroid belts don't. In reality, asteroids are spread over such a high volume that it would be highly improbable even to pass close to a random asteroid.

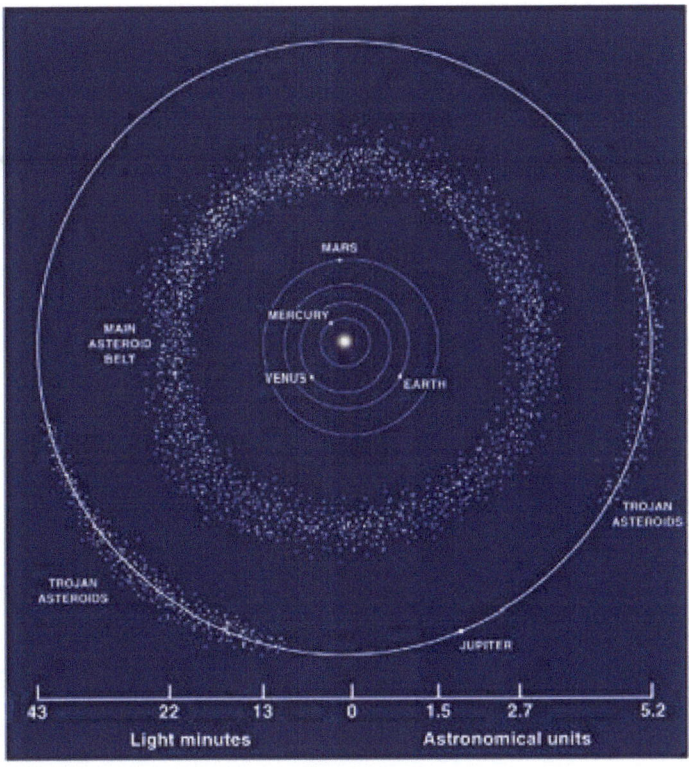

Above The Earth

It is easy to forget that the human family lives on a small blue planet named Earth. All around us we see trees, animals, cars, buildings, farms, factories, stores, and other natural and man-made structures.

With all of these every day familiar objects around us and with the blue sky above us, and the deep oceans beneath us, our home planet often feels quite large. Compared to us, it is very large. There is enough space for each of us, our families and friends, our pets, as well as trillions of other life forms to live and enjoy the various experiences of life.

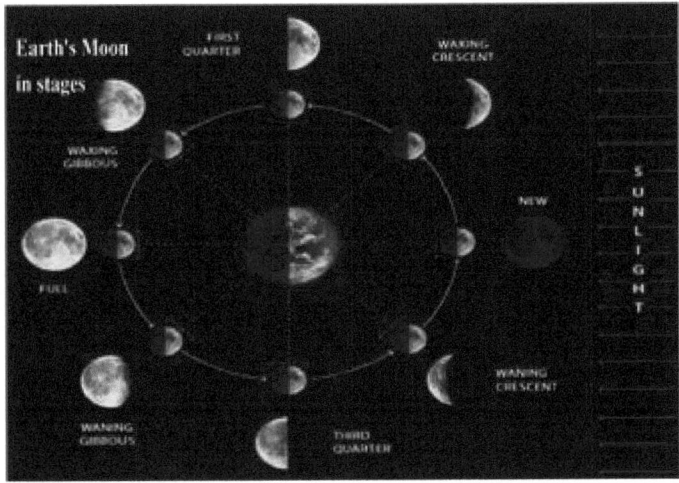

The Moon is like a desert with plaints, mountains, and vallies. It also has many craters, which are holes created when space objects hit the Moon's surface at a high speed. There is no air to breathe on the Moon. Not long ago frozen water was discovered at each pole; the top and bottom of the Moon. Scientists think ice was left over from a comet that once collided with the Moon.

The Moon travels around the Earth in an oval shaped orbit. Scientists think the Moon was formed long ago when Earth collided with another space object. That collision may have caused a big chunk of rocky material to be thrown out into space to form the Moon. The Moon is a little lopsided. Its crust is thicker on one side than the other. The Moon is much smaller than the Earth. The pull of its gravity can still affect the Earth's ocean tides. We always see the same side of the Moon from Earth. You have to go into space to see the other side.

The Moon

Luna, or simply 'the Moon', is a relatively large terrestrial planet-like satellite, about one quarter of Earth's diameter. The natural satellites orbiting other planets are called "moons", after Earth's Moon.

While there are only two basic types of regions on the Moon's surface, there are many interesting surface features such as craters, mountain ranges, riles, and lava plains. The structure of the Moon's interior is more difficult to study. The Moon's top layer is a rocky solid, perhaps 800 Km thick. Beneath this layer is a partially molten zone. Although it is not known for certain, many lunar geologists believe the Moon may have a small iron core, even though the Moon has no magnetic field. By studying the Moon's surface and interior, geologists can learn about the Moon's geological history and its formation.

There is no wind on the Moon. The Moon does not possess any atmosphere, so there is no weather as we are used to on Earth. Because there is no atmosphere to trap heat, the temperatures on the Moon are extreme, ranging from 100° C at noon to -173° C at night.

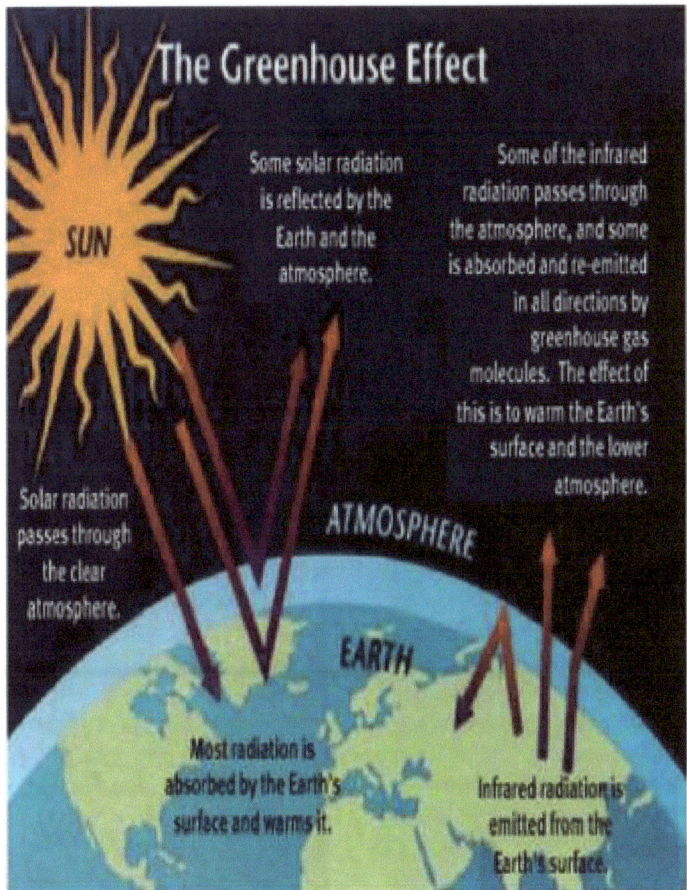

Natural and environmental hazards

Large areas are subject to extreme weather such as tropical cyclones, hurricanes, or typhoons that dominate life in those areas. Many places are subject to earthquakes, landslides, tsunamis, volcanic eruptions, tornadoes, sinkholes, blizzards, floods, droughts, and other calamities and disasters.

Above The Earth

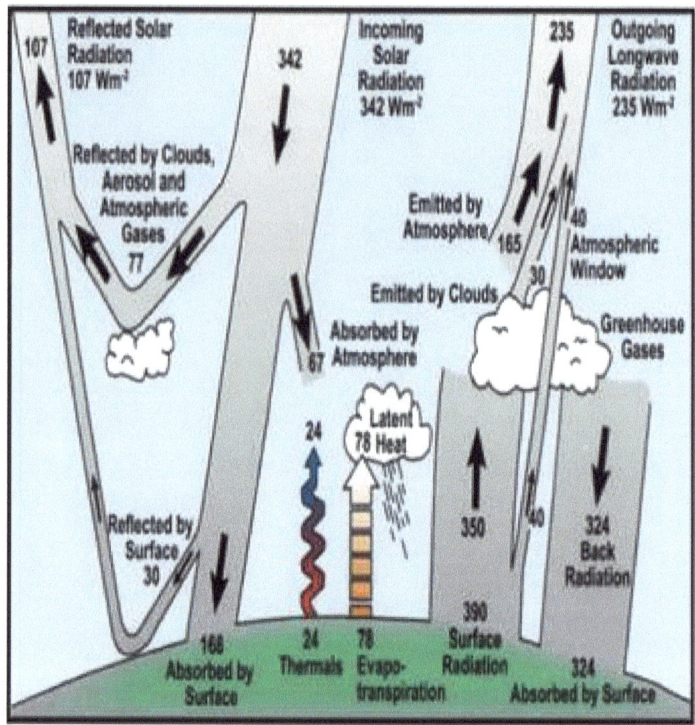

Factors that affect climate on Earth

Many localized areas are subject to human-made pollution of the air and water, acid rain and toxic substances, loss of vegetation, loss of wildlife, species extinction, soil degradation, soil depletion, erosion, and introduction of invasive species.

A scientific consensus exists linking human activities to global warming due to industrial carbon dioxide emissions. This is predicted to produce changes such as the melting of glaciers and ice sheets, more extreme temperature ranges, significant changes in weather conditions and a global rise in average sea levels.

Liberty Dendron

Modern geologists and geophysicists accept that the age of the Earth is around 4.54 billion years (4.54 × 109 years ± 1%). This age has been determined by radiometric age dating of meteorite material and is consistent with the ages of the oldest known terrestrial and lunar samples.

Following the scientific revolution and the development of radiometric age dating, measurements of lead in uranium-rich minerals showed that some were in excess of a billion years old. The oldest such minerals analyzed to date, small crystals of zircon from the JackHills of Western Australia, are at least 4.404 billion years old. Comparing the mass and luminosity of the Sun to the multitudes of other stars, it appears that the solar system cannot be much older than those rocks.

Rich in calcium and aluminium, the oldest known solid constituents within meteorites that are formed within the solar system , are 4.567 billion years old, giving an age for the solar system and an upper limit for the age of Earth. It is hypothesized that the accretion of Earth began soon after the formation of the Ca-Al-rich inclusions and the meteorites. Because the exact accretion time of Earth is not yet known, and the predictions from different accretion models range from a few millions up to about 100 million years, the exact age of Earth is difficult to determine. It is also difficult to determine the exact age of the oldest rocks on Earth, exposed at the surface, as they are aggregates of minerals of possibly different ages. The Acasta Gneiss of Northern Canada may be the oldest known exposed crustal rock.

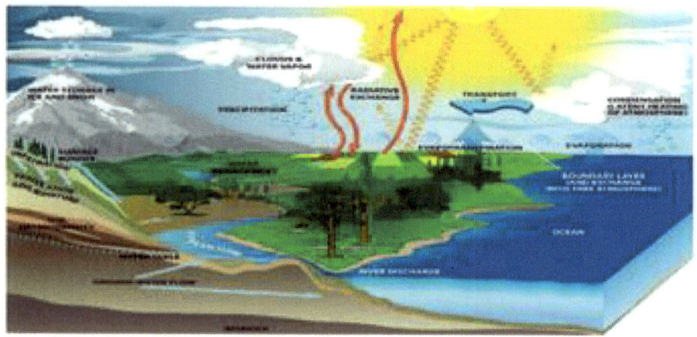

Earth in the Solar System

It takes Earth 23 hours, 56 minutes and 4.091 seconds to rotate around the axis connecting the north pole and the south pole. Earth orbits the Sun every 365.2564 mean solar days. Earth has one natural satellite, the Moon, which orbits around Earth every 27 1/3 days.

The orbital and axial planes are not precisely aligned: Earth's axis is tilted some 23.5 degrees against the Earth-Sun plane (which causes the seasons), and the Earth-Moon plane is tilted about 5 degrees against the Earth-Sun plane, otherwise there would be an eclipse every month.

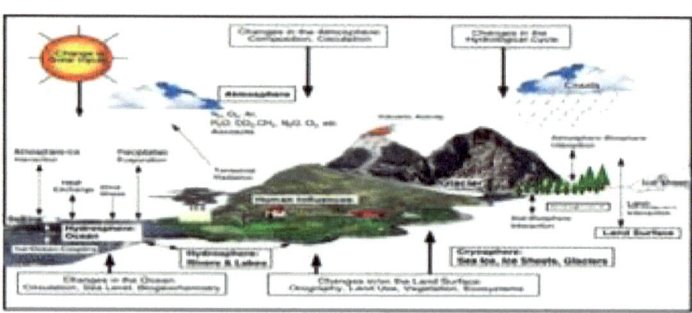

Earth in the Solar System

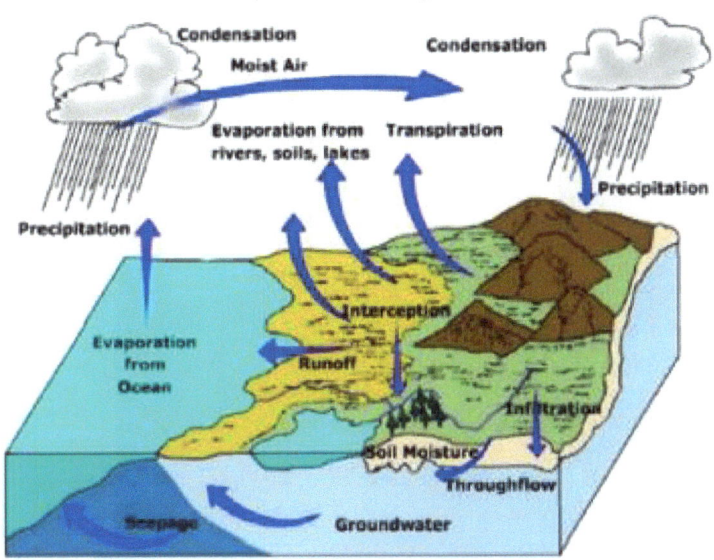

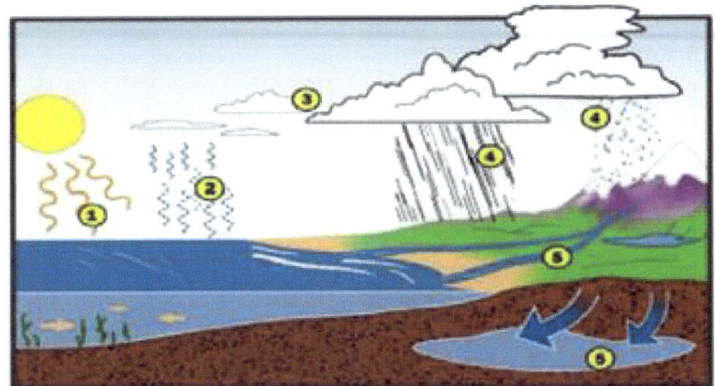

1. The sun heats the ocean.
2. Ocean water evaporates and rises into the air.
3. The water vapor cools and condenses to become droplets, which form clouds.
4. If enough water condenses, the drops become heavy enough to fall to the ground as rain and snow.
5. Some rain collects in groundwells. The rest flows through rivers back into the ocean.

Above The Earth

Lafayette is....Liberty Dendron. He was born in Norfolk Virginia. Attended Norfolk State College, and graduated from Adelphi Business School. He has a Bachelor of Arts degree in Creative Writing, from Glendale University. He lives in Dendron Virginia.

www.ingramcontent.com/pod-product-compliance
Lightning Source LLC
Chambersburg PA
CBHW041143180526
45159CB00002BB/720